The Four Seasons

Gardening Journal

Christine Dunne

Christine Dunne, Publisher

Salinas, California 2020

ISBN-978-1-7350162-0-7

Printed by Lulu Press, Inc. in the United States of America.

First Printing, 2020

Christine Dunne, Publisher

P.O. Box 2002

Salinas, California 93902

www.deadland.co